Palewell Press

Tutu's Rainbow World
Selected Poems

Joseph Kaifala

Palewell Press

migrations

Tutu's Rainbow World – selected poems

Published by Palewell Press Ltd
http://www.palewellpress.co.uk/

First Edition

ISBN 978-0-995535-18-3

Cover design Copyright © Camilla Reeve 2017
Cover image courtesy of the wonderful image library at https://www.pixabay.com
A CIP catalogue record for this title is available from the British Library.

Palewell Press Ltd supports the Forest Stewardship Council® (FSC®) the leading international forest-certification organisation. Our books carrying the FSC® label are printed on FSC®-certified paper. Their printing and binding complies with ISO 14001 (Environmental Management) and 50001 (Energy Management).

Dedication

Wole Soyinka gave Nelson Mandela the earth. I give Archbishop Desmond Tutu the world that he devoted his life to painting in rainbow colors.

To Desmond Mpilo Tutu
(For his courage and strength)

Introduction

If, as Chinua Achebe wrote, "proverbs are the palmoil with which words are eaten," then poetry is the *lafidi* (condiment) that makes words tasty. I am rooted in ancient traditions of griots and town criers whose language is poetry. The message and the manner in which it is conveyed are equally important. In a few lines, one may communicate what others might need three hundred pages to express. It is this ability to conjure, in a verse, a universe others can visualize and experience that garnered my interest in poetry.

I was also lucky to grow up in the company of my late uncle, Mohamed Lord Vandi Tongi, for whom poetry was the greatest art. I remember his characteristic long pauses between lines and verses to emphasize things he wanted me to notice, hear and feel. I never forget the melancholic expression on his face as he repeated the thirteenth line of his favorite poem, *The Vultures* by David Diop: *you who knew all the books but knew not love*. I now understand why this line made my uncle pause longer. Education devoid of love is inadequate.

Moreover, I grew up in circumstances for which I believe poetry is the most appropriate language. I lived through the Liberian and Sierra Leonean civil wars and inadvertently became a living witness. I write so that others may glance at the things I saw and perhaps together we can make the world better than it is.

I am appalled by the present lack of concrete solutions to the Syrian conflict and the use of children as combatants. I believe that with honest approach and leadership we can end the conflicts in Somalia, Democratic Republic of Congo, Israel-Palestine, Myanmar, and other places of perpetual upheavals. The immigration crisis in Europe is a reminder that we must address the problems that force people to flee their homes, making perilous journeys, for a chance to survive. Many of the problems that limit human progress are artificial. Therefore, responsible leadership is our greatest solution.

I am grateful for the support of my friends and family, especially Rose Tolno, Prof. Catherine Bertini, Mousoukoro Sandouno, Mariata Joe Dioubate, Jonathan Lobbo, Marie Jeanne Leno Kerian, Ryan Kerian, Jeanne Keifa, Fr. Edwin Turay, Abdo Assad, James Omo Oloye, Ida Rose Nininger, Dunia Bah, Dorsave Ben Salah, Jess Wilkerson, Eliza Paul, Leonard Gordon, Sadiatu Marrah, Zainab Moserey, Fatmata Kamara, Mrs. Josephine Kamara, Mr. Alusine Kamara, Florence Carew, the Bainda family, Augustine Bendu, Jonathan Edwards, Kipsy Ndwandwe and my SkidFam siblings. My uncles, Sahr Joseph Tolno, Joseph Lobbo, Alia Sekou Sandouno, George Sandouno, Tamba Keifa, Adam, Patrick and Isaac Tongi. My aunts, Marian Yillia, Elizabeth and Marie Louise Kantambadouno. Many thanks to Camilla Reeve and Palewell Press.

Contents

Tutu's World

He carried the weight of Africa
affixed to his shoulders
like a hunchback on an aged back.
Shielded by his vest of holy cassock
he raged on the battlefield for freedom
and, in the pulpit, he shepherded his children
toward the colorful beam of a rainbow nation.

There is no room for hate in Tutu's world,
only wayward children in need of God's love.
An unusual knight in holy garment,
he fought to rescue man from man.
In his blessed purple armor
he stood fearlessly on the frontline,
a cross-wielding Spartan
defending his flock from their own evil hands.

He took no eye for an eye
but painted a rusty world
with flamboyant love.
While lieutenants of the struggle
languished in apartheid dungeons
he sprinkled holy water on saints and sinners
to wash away the tribulations
and cruelty their eyes had seen.

/continued

1

When the battlefields were silent
and casualties lay wasted in the ashes of hatred,
like a neutral nurse of a red emblem mission
he gathered the dead and wounded
and purified the land of blood and curse.
Then, with his supplicating hands,
sanctified the earth from lingering ghosts,

that we may forever live in peace,
in Tutu's rainbow world.

Move the World

One more step along the world I've come,
along pavements that seem to lead beyond,
from familiar stories, a few years done,
to a new page almost unknown,
landing on shores that may be holding
philosophies different from my own.

From the lands of the Bambara
where Yemani trees stretch afar,
I have swum across the Mano River,
uniting fallen tribes of separate lands,
bearing once more renewed
the humble peace of my forefathers.

If these whirling winds can offer help,
I shall arise from the lion's den,
passing through the old and crafting new,
the cosmos a shape that's yet undone,
to leave my mark where I belong.

Lion Mountains

In the Lion Mountains, I was conceived.
Absent from there, I have redefined myself.
I was compelled to leave you, motherland,
but I am strengthened by your blood,
the blood of free slaves mixed with that of natives
on the western side of a continent,
a home for Africa's abandoned children.

How I long to coil again in your black belly,
smell the aroma of *fufu* and *tola,*
cooked with *hog-foot* and *canya pepe,*
the tang of Mama Jeneba's *pemahun*
mingled with that of *kenda* and *dry-fish*;
to fold my index finger and lick the *masangé*
that made me a strong and healthy boy.

Land that we love, our Sierra Leone,
where lunch was never salad or a sandwich
but fine *bo gari* and a slice of *kanya* à la carte,
where dinner was served at the bottom of a pot,
numerous hands competing to feed each mouth;
the children left to scrape the *krawo,*
since hands and ages don't play fair.

Land of my humble birth where I belong,
I weep for you in the silence of my night,
knowing your children still struggle for their life.
As you attempt to rise from the ruins of war,
slowly, painfully, your head penetrates the rubble,
hope and your children pulling you up
in witness to your zeal that never tires.[1]

---

[1] Fufu: Dough made from fermented cassava
Tola: Slippery plant powder used to make fufu sauce
Canya pepe: Cayenne pepper
Pemahun: a potpourri of rice and potato leaves with condiment
Kenda: A West African condiment with a pungent smell
Masangé: a type of palmoil
Bo Gari: dried ground cassava
Kanya: a mixture of rice powder and peanut butter
Krawo: crust made by cooked rice

I weep at night

I weep at night
when the noise of day
dissolves to silence,
when the moon overthrows the sun,
the town falls into decorum
and my world slumbers
under the cover of darkness.

I weep at night
when creepy creatures crawl from their burrows.
Somewhere around the world
others awake to the chaos of day
while my pillow drowns in tears.
I regret another day gone by
that left no legacy.

I weep at night,
having woken the day before
with plans of saving the world,
to realize I'm in bed once more
with the world the way it used to be—
nothing but the same old place
where I must learn to survive.

I weep at night
knowing Palestine still struggles,
in Fallujah, there's no time to sleep
and Syria crumbles to dust.

Cursed by the inability to change things
I've no more tears to shed.
For South Sudan, I shall lie in vigil.

I weep at night,
knowing that if I do not,
there's no-one who will mourn
the faceless children of Mogadishu
and those who perished before them—
in the way we now deem normal—
the hazards of human existence.

I weep at night
as a worried mother, somewhere,
watches over her sleeping child,
knowing the world as a peril
where her child must grow.
She cannot rest at night
but listens for the echoes of dawn.

I weep at night
as broken promises haunt my soul.
I came to save the world
but it's I who need saving now
My days have come to a close,
soaked in the dew of death.
In Al Jannah, I shall rest.[2]

---

[2] Jannah: Paradise in Islam

Encounter with Love

I encountered love just once.
She appeared without mystery
and offered herself to me.
I was too foolish to recognize her,
wrapped up in myself.
I couldn't feel the warmth of her palms
or hear the soft whisper behind my ear
that said, "I've been here all along."
She walked away, and never looked back.
I watched her disappear,
her shadow dissolving with her gait
like an apparition from beyond the clouds.
In the chills of her departure,
I woke to the realization
of her reluctant goodbye,
and in the agony of that knowledge
it became all too clear—
love would never return to me.

Hope

Hope brings us through night
to the crack of dawn.
Cock crows the morning
at the break of day.
Even for those without plan or chore,
hope brings courage at dawn.
To those less fortunate,
fall flowers blossoming in spring,
you too may smile—
Someday, stars even shine
on those who start from the bottom.

Sunday Morning in my Home Town

Sunday morning in my hometown.
Mother is busy making our morning meal
while papa irons our fancy clothes.
Mother does not allow sister to iron clothes
anymore—
several Sundays ago, sister burned mother's
favorite *lappa*.
This morning, unlike weekdays, we children must
bathe.
"Cleanliness is next to godliness," I hear mother
say.
That is why we bathe properly on Sundays
instead of our usual water-to-belly ablution.

Cathedral bells are ringing in the distance.
The neighborhood churches are beating their *bata*,
raising a cacophony of calls to Jesus.
Our neighbors, the Born-Agains, have already left
for church.
It is only their second service of the day—
Earlier, they attended a 5 a.m. worship,
footsteps audible beneath the *cocoriocos* of
roosters.
There's one more evening prayer before their
Sunday ends.

Our Catholic mass lasts an hour and a half
to which half the congregation comes late.
Papa would love to be among the late churchgoers,
but mother will not miss her hour of rosary
and we children must attend Sunday School.

Between the Responsorial Psalm and Second
Reading
that's when half the congregants are ushered in.
Our European priest can do nothing at all
to make his African flock come to the Lord on time.
He must accept that, here in Freetown,
God, like man, shall always have to wait.[3]

---

[3] Lappa: piece of cloth tied around the waist
Bata: drum
Cocoriocos: African onomatopeia for cock crow

God Was There

God was there
when our guns clattered,
our rice fields were engulfed in fire
and our dwellings sank in flames;
when brothers bayoneted sisters,
spilling blood in sacred places;
our languages were sullied with hate
and our customs crumbled
to the rhythmic ratatat of war.

God was there
when children were molded
into unwilling implements of death,
walking zombies pillaging the earth
that brought them into being,
unrelentingly shattering their future,
these monstrous infants with killing tools
entrapped in the lunacy of a power struggle
in which they had no share.

God was there
when we shed tears
like mourners at the death of tribal chiefs
but we were mourners at our own graves
dug by our bloodstained hands
beneath the rubble of a time
before we had abandoned
the teachings of our ancestors,
becoming hunters of ourselves.

God was there
when bloodthirsty monsters
arrived to tear us to pieces,
stretched from both ends
like the rubber strings of a catapult.
Unable to draw any further,
they let us fall apart like a broken necklace;
and God, he was there all along,
a bystander in the midst of death.

Harmattan

(For Leonard)

I wake up
in the chilly breeze
of a harmattan morning,
skin covered in ash
like a man just resurrected.

All around my neighborhood
small crowds gather around flames
in ritual gratitude
to the god of fire
whose gift of warmth
repels the harmattan cold.

Soon heat will defeat cold
as burning wood becomes cinder
and sweaters slide off bodies
in search of some coolness
beneath a mango tree.

As the sun hides behind dust clouds
and the frigid harmattan returns
mothers tell their children
not to bathe in the river,
there is enough dry wood
to heat their bathing waters,
they must wear their sweaters again
against the harmattan cold.[4]

---

[4] Harmattan - a very dry, dusty easterly or north-easterly wind on the West African coast, occurring from December to February.

Love

(For Chernor and Aissatou)

Love,
in whose tender hands
our fate is sealed
when our fragile hearts
in their individual cages
beat only in partial rhythm
but together make a perfect harmony
to keep us dancing,
body to body,
heartbeat by heartbeat
in an everlasting Kizomba of life.

Never Again

We went astray
and forgot briefly
who we were.
When our light was rekindled
we saw before us
a wreckage of our old selves
and we promised never to forget
ever again.

Freetown Cotton Tree

Roots spread
like an octopus
grabbing the earth,
like claws
in tender flesh,
veins penetrating the soil
like catheters.

The Cotton Tree sits
like an ancient deity
on her golden stool.
Age has done nothing
to her protective presence,
like a grandmother
watching vigilantly
over her grandchildren.

Years of nurturing her own children
have not diminished her love
for her children's children
who walk at her feet, daily,
as they perambulate
in pursuit of life's offerings
in the heart of Freetown,
a city always protected
by her shadow
and the aged presence
of her giant stem.

God's Piece of Earth

Young as a cob,
standing with a Kalashnikov
against a dilapidated wall
staring at unexploded missiles
dropped the night before,
or maybe years ago.

His twisted face reveals
the weight of his Kalashnikov
attached to his tiny body
by an unbreakable cord.
The pain in his young eyes
says he is bound to carry guns.

Young as a cob,
snatched from his mother's milk
to bear the cold burden of war,
a call to duty that cannot wait,
duty for every boy
to become a soldier.

From birth, his mission defined—
to kill in the name of God.
For him there's but one call—
die fighting or get killed.
A warrior's death makes martyrs of men.
A peaceful death, eternal damnation.

/continued

So, he fights and gets killed.
Another mother wails tonight.
Borne on the shoulders of men
he is hailed as a martyr
to lure others of his age to fight.

Rest in peace, God's Piece of Earth.

Servants of the Earth

Indentured servants of the earth,
love it as a whole
to let it serve all as one
and take only as need be
what blessings the earth offers
to all of humanity.

Do you remember those days?

Do you remember those days
when love reigned in our hearts?
I danced to melodic country songs
while you cooked our vegetarian meal,
smiling at my disorganized flinging of twigs
and my discordant attempts to sing.

Do you remember those days
when I ate your food with diligence?
Like a cow grazing on fertile grass,
I ate slowly and with serene devotion,
watching the smile upon your glowing face,
your pride in nourishing my soul.

Do you remember those days
when love flourished in our hearts?
Like potato leaves in the rainy season,
we walked hand-in-hand,
linked by the cord of our affection
in the darkness of warm coastal nights.

Do you remember those days
when I held you close to me
as we walked down Franklyn Street
reciting stories of our day,
finding common ground between Arabian folklore
and dismal tales of nuclear doom?

Do you remember those days
when I pulled you around Alvarado Street,
consoling you with Shakespearian monologues,
you smiling forgiveness at my folly?
We walked slowly and quietly in dim lights
pulling each other from side to side like playmates.

Do you remember those days
when passers-by smiled in approval
as love pulled us along like puppies,
window-shopping outside neon-lit stores,
energized by salty Pacific breezes
and the aroma of Little India cuisine?

Do you remember those days
when we strolled through Monterey Park
consecrated by herbal fragrance?
As we merged with Scott Street
covered by the shadows of trees
I kissed you quietly.

Do you remember those days
when we reemerged on Watson Street,
two deer always waiting there for us?
You paused and watched them closely
as if to seek the meaning of love
in the friendly couple's eyes.

Do you remember those days
my love and my darling,
when we lay down on my sofa
and I kissed you softly
throughout the summer night
in the place our love was conceived?

Lampedusa - my brothers' grave

Lampedusa—my brothers' grave.
From rickety slums and human misery
my brothers and sisters have come,
tangled in the bond of awful memories,
confined in the holds of their wretched vessels,
huddling and cuddling in their wooden graves,
the hope of a better life, their only lighthouse.
Where they perish, others will follow.

Lampedusa—my brothers' grave,
where tombstones float on salty liquid,
numbering not in single file
but in hundreds of liquid mass graves.
Ocean sometimes rejecting their black bodies,
bloated corpses washed ashore,
some nibbled by indifferent sea creatures.
Here lies Africa's future, embalmed with salt.

Lampedusa—my brothers' grave.
Where Africa's children cast their last lots,
battling treacherous waves upon a vast ocean,
these black pirates have only hope for booty,
hope shattered by their individual nations.
They cast themselves upon the sea
for a better life on another's shore,
made for themselves a final resting place.

Lampedusa—my brothers' grave.
In their hundreds, they keep coming.
Like saifu ants, they file to their end.
Even where news of death abounds
this their fate that Africa has tolled,
no sepulcher to remember their courage.
In Lampedusa, may they rest in peace.[5]

---

[5] Since 1998, the tiny Italian island of Lampedusa
has been one of the main European entry point for
immigrants from Africa. The hulks of fishing boats,
peppered with gaping holes, can be seen all over the
island.

Love Comes

(For Eliza)

Love comes at night
to seize my gentle heart
while nightingales chant
hymns of adulation in the distance
reviving memories of you,
where you were once
a loving part of me.

Where you slept
your fragrance now awakens
to remind me of your presence,
a beauty I cannot embrace
even as I inhale your scent,
believing you are here,
sleeping right beside me

There was a meeting in Heaven last night

Hear ye, Hear ye!
Did the town crier not pass through your town?
There was a meeting in Heaven last night.
What meeting do you speak of, Ear of the Dead?
Chief Albert Luthuli was there.

I said, what meeting do you speak of, Messenger of
our Ancestors?
Oliver Tambo was in attendance and he sat with
Mandela.
Chris Hani was there, too, and he sat at the feet of
Old Luthuli.
Ruth First and Joe Slovo were there.
Robert Sobukwe sat with his chin in his palm.
Then the child Hector Pieterson came in and sat
with Ruth.

What was said at the meeting, dear wise one?
Chief Luthuli cleared his throat but his mouth
remained shut.
Comrade Oliver unfolded his arms in an attempt to
speak
but he leaned back and no words were spoken
Young Hani raised his hand but quickly brought it
down.
Joe and Ruth sat next to each other with their arms
folded.
All along Mandela sat with a mischievous smile.

/continued

What stilled the tongues of those roaring lions, only
the gods know!
At one point they all looked down toward South
Africa.
Just then, an Angel of God passed by.
He mocked them saying: "That is Mandela's Earth!"
There was silence. Then Mandela spoke:
"There, they desecrate the land we gave them to live
free!
We carried their crosses, now they spill the blood of
their brothers and sisters.
How easily they forget the battle Africa fought when
no one else would!
The Spear of the Nation that gave them freedom,
they now use to pierce Africa's side.
It is the greatest betrayal of everything we built on
that southern edge of a continent!"

Hear ye, Hear ye!
After Mandela spoke, the room was once again
silent
Then, crying, the Old Chief sang, *Ngosi sikelel
iAfrica*.
Oliver, hearing his elder weep, also cried.
A pool of water flowed down young Hani's face.
Our Mandela was smiling again and appeared
untroubled
but after a while, he too had tears dripping down
his cheeks.
They wept!

When they'd wiped their eyes and rubbed their
noses
Mandela mumbled something the Old Chief
couldn't hear.-"What did Comrade Mandela say?"
he asked young Hani,
who sat leaning on the old man's right leg.
"He is worried about Comrades Tutu, Kathy, and
others!"
Looking down, they saw their living comrades and
pitied them.
Kathy and Tutu were huddled together in silence,
shivering with goosebumps from the massacres at
their door.
The Old Chief's heart became heavy with grief,
He rose and wandered off slowly, the others
followed,
leaving the room where they had sat for many
hours.
So ended the meeting in Heaven last night!

Woman of Africa

(For Sadia & Kipsy)

African woman,
woman of the north,
woman of the south,
woman of the east and west,
woman of the Makona river,
how beautiful you are

I look at you without blinking,
your black braided hair
a beauty I cannot resist
like an alignment of the stars,
your smooth dark skin
complements your fierce white eyes.

I could stare at you forever
as you do your majestic walk,
feet and hips in rhythm
to the beat of your internal *dundunba*.
With the attitude of a goddess
you flirt with my unblinking eyes

Woman of the fields,
woman of the rivers of Africa,
I feel your strength within me.
Yours was love at first sight.
I inhaled your fragrance
and became man.

Woman of Africa,
bearer of the black race,
beautiful is your voice
that echoes across a dark continent,
blooming in the hearts of men
who long to hold you.[6]

---

[6] Dundunba: drum

Child Soldier

How I miss those infant days
when I was just a child
on my mother's back,
carried with tenderness
while I slept,
not knowing the world
outside my mother's lappa.

How I wish I knew
the reason I now bear
this gun that weighs me down,
my only possession.
I kill to preserve my life.
What kind of world is this
where a child must kill to live?

How does one choose to live
in a world such as this
where children fight in wars,
rape when they know not
how to give love,
take life when they know not
how to give life?

The agony of sobriety
when the drugs run their course,
after they'd woken the killer in me,
to shoot at will
whoever stood in my path.
I gorged on their eyes
and ate their hearts,

I live on the life of my enemies—
other children my own age
intoxicated by the labyrinth of war
whose only choice is to die.
I languish with the hope
to live and die another day
of childhood.

School

Softly I exchange the fine smooth surface of my
Congressional pen
for the rough wooden scales of a laborer's pickaxe
to construct a fountain of knowledge for my
people—
in generations to come
pens shall give birth to machinery
and the children of this land
shall excavate with pens
the abundant gifts a fertile mind can bring.

King Jimmy Market

No-one who lives in Freetown
will ever in good faith wonder
what in heaven's name
goes down at King Jimmy market;
where hawkers, merchants, and criminals
trade wares of questionable origin,
some the product of hard work and sweat,
others stolen from the neighbor's yard.

When one cannot find
what one seeks elsewhere,
at King Jimmy, the item awaits,
either in the kiosk of an honest trader
or in the backpack of a *ju-man*.
No questions are ever asked at King Jimmy
just as no merchandize carries a fixed price.
Whatever one needs one finds
in the bazaar of King Jimmy.[7]

---

[7] Ju-man: hustler

Cackle

Laugh a good laugh,
expand the facial curvatures
into a canvas of joy,
flatten the undulations of pain
in defiance of knotted ends,

the curse of pain and sorrow
conquered with careless leverage,
a remedy for those who languish
beneath the yoke of life's challenges
Like a bird caught in a hurricane
seeking refuge on a pole.

If a pole is all there is at present
don't think life has no glories,
let your jaws shift and cackle
like the laughter of an idle witch
in the silence of her mystic chamber.

Let your sweet tears roll
to the beat of your heart.
Exorcise the demons that clench your lips,
allowing your vibrant laughter
to subdue the most potent sorrows.

Laugh a good laugh
and let the coolness of your tears
refresh your weary soul
and infect those who may be altogether
resigned to enduring pain.

Birthday

(For Liat)

Born to grow
like an unfolding flora
with no compulsion to life
until your rosy petals bloom
and excited bees ruminate for nectar
around the sweet fluid of your being.

From the winter of your birth
the spring of life is born.
For those waiting patiently in life
joy unfolds on the first day of a year.
For they who eagerly await
the beginning of an epoch,
a child like you is born.

Of Africa

Amidst the tragedies of Africa, I celebrate
Egypt,
the land of Khnum Khufwy
where the pyramids remain an unbeatable
landmark.
Tanzania gives us the beautiful spread of the
Serengeti
beneath Mount Kilimanjaro.
In Kenya, giraffes still stand tall
and human feet stretch on racetracks like cheetahs.
Ghana and Botswana—
where our aspirations are germinating.
I say thanks to Nigeria for sages and scribes
who chronicle our stories.
Dearest Mali, where Timbuktu stands,
and griots continue to chant our hymns of hope.
I raise my glass to South Africa and say
Amandla! Amandla! Amandla!
For you gave us Mandela and Tutu,
the stars that shine from the south.

The Walk

She knows I am watching
and so she walks the walk
that rewards my interested eyes
but I refuse to accept her charge
and look the other way.
As she disappears down the road
I catch myself watching again.

The Gods Have Willed

The gods have willed
to those who dare listen
these proclamations on savior-vivre:
among the dark clouds of life,
a divine light brightens the way
toward glorious stars of peace.
Though we insatiably yearn for abundance
like tilapia gulping on flowing waters,
what we take we do not consume
but let it flow again
from nothing to nothing.
Each moment in life throws us further,
sometimes harder than before.
Like branches we dance in the wind—
some break and others sway to stand again.
When, in good weather, the waters are calm
we sail smoothly to shore
as captains of our own vessels.
We stand amidst the gods like angels
yet we know not where our powers lie.
We wandered too far without fear
but now we must stay within the boundaries of love,
where the cardinal points meet at the center,
piercing the core of our hearts
and stitching us together,
like the two sides of a kola nut.

Mother Came Home

Mother came home
when the sun set at dusk
bringing chaos to peaceful quarters.
As the day's work in the fields
at dusk gives life to town,
the cold ashes of fireplaces
produce flames of evening fire,
smoke rising like rockets,
the aroma of palm-oil and cassava leaves
offers hope to hungry children.

Mother came home tonight
but palm trees no longer swing
to the soft touch of evening breezes
that spread the aroma of palm kernel oil
from ye Kadie's kitchen,[8]
the origin of savory smells.
And the voices of cheerful children
no longer echo above the town.
The fireplaces are cold again

---

[8] Ye: Mother (Mende)

Tutu's Rainbow World

Joseph Kaifala - Biography

Joseph Kaifala, Esq. was born in Sierra Leone and spent his early childhood in Liberia and Guinea. He later moved to Norway where he studied for the International Baccalaureate (IB) at the Red Cross Nordic United World College before enrolling at Skidmore College in upstate New York. Joseph was an International Affairs & French Major, with a minor in Law & Society. He holds a Master's degree in International Relations from the Maxwell School at Syracuse University, a Diploma in Intercultural Encounters from the Helsinki Summer School, and a Certificate in Professional French administered by the French Chamber of Commerce.

Joseph was an Applied Human Rights fellow at Vermont Law School, where he completed his JD and Certificate in International & Comparative Law. He is recipient of the Skidmore College Palamountain Prose Award, Skidmore College Thoroughbred Award, Vermont Law School (SBA) Student Pro Bono Award, a 2013 American Society of International Law Helton fellow, and a member of "Who Is Who Among Students in American Universities & Colleges" in recognition of outstanding merit and accomplishments as a student at Vermont Law School. Joseph was one of the BBC World Service Outlook Inspirations Fifteen. As the programme described it, these are "people who show us a better side of being human."

The Jeneba Project

The Jeneba Project Inc. is a charitable organization [US 501(C)(3)] devoted to providing educational opportunities to the children of Sierra Leone. The organization is currently fundraising to complete the construction of a secondary school in Robis, Northern Sierra Leone. Learn more at www.jenebaproject.org

Palewell Press

Palewell Press is an independent publisher handling poetry, fiction and non-fiction with a focus on human rights, social history and the environment. The Editor may be reached at enquiries@palewellpress.co.uk

www.ingramcontent.com/pod-product-compliance
Lightning Source LLC
Chambersburg PA
CBHW071126030426
42336CB00013BA/2221